Department of Commerce and Labor

BUREAU OF NAVIGATION

RULES AND REGULATIONS

RELATING TO THE

ANCHORAGE OF VESSELS IN THE PORT OF NEW YORK

APRIL 25, 1907

WASHINGTON
GOVERNMENT PRINTING OFFICE
1907

1837
ARTES
SCIENTIA
VERITAS
LIBRARY OF THE
UNIVERSITY OF MICHIGAN
SI QUAERIS PENINSULAM AMOENAM
CIRCUMSPICE

Department of Commerce and Labor

U. S. BUREAU OF NAVIGATION

RULES AND REGULATIONS

RELATING TO THE

ANCHORAGE OF VESSELS IN THE PORT OF NEW YORK

APRIL 25, 1907

WASHINGTON
GOVERNMENT PRINTING OFFICE
1907

AN ACT RELATING TO THE ANCHORAGE OF VESSELS IN THE PORT OF NEW YORK.

Be it enacted by the Senate and House of Representatives of the United States of America in Congress assembled, That the Secretary of the Treasury [Commerce and Labor] is authorized, empowered, and directed to define and establish an anchorage ground for vessels in the bay and harbor of New York and in the Hudson and East rivers, to adopt suitable rules and regulations in relation thereto, and to take all necessary measures for the proper enforcement of such rules and regulations.

SEC. 2. That in the event of the violation of any such rules and regulations by the owner, master or person in charge of any vessel, such owner, master or person in charge of such vessel shall be liable to a penalty of one hundred dollars; and the said vessel may be holden for the payment of such penalty and may be seized and proceeded against summarily by libel for the recovery of the same in any United States district court for the district within which such vessel may be, and in the name of the officer designated by the Secretary of the Treasury [Commerce and Labor].

SEC. 3. That this act shall take effect immediately.

Approved May 16, 1888.

RULES AND REGULATIONS GOVERNING THE ANCHORAGE OF VESSELS IN THE PORT OF NEW YORK.

DEPARTMENT OF COMMERCE AND LABOR,
OFFICE OF THE SECRETARY,
Washington, April 25, 1907.

The following described anchorage grounds for vessels in the bay and harbor of New York, and in the Hudson and East rivers, respectively, are hereby defined and established, and the following revised rules and regulations governing the same are published for the government of the owner, master, pilot, or other persons having charge of a steam vessel towing or otherwise conducting a vessel to an anchorage in the port of New York, pursuant to the act of Congress approved May 16, 1888:

(*a*) Vessels shall anchor only within the following specified limits, and a vessel anchoring within any of the below described boundaries must anchor entirely within said boundaries, so that no portion of the hull shall extend beyond said boundary after veering chain or when riding to a tideway.

(*b*) No vessel shall anchor in any of the channels except in cases of great emergency, and then as near the edge of the channel as possible, so as not to impede or interfere with the free navigation of the same, and only until such time as they can procure assistance; and no vessel shall anchor so as to obstruct the approach to any pier or impede the movement of any ferryboat.

(*c*) All stakeboats used for assembling barges, canal boats, and other vessels preparatory to being made up in tows, and vessels used for storing explosives and moored only in the anchorage for explosives, as described in paragraph 2, No. 27, page 9, may be moored under permits granted by the supervisor of anchorages and moored only in such places as he may designate.

(*d*) A vessel upon being notified to move into the anchorage limits must at once get under way or make a signal for a tug.

(*e*) Permits may be granted by the supervisor of anchorages to wrecking plants to anchor in the channel for the purpose of recovering sunken property, subject to his supervision. Such wrecking plants must comply with all the navigation laws in regard to lights, fog signals, etc., and in granting such permit the Government assumes no responsibility.

(*f*) Points where cables and pipes cross the anchorage grounds are marked in red on the accompanying maps, and all vessels are cautioned not to anchor so as to interfere with them.

(*g*) All ash scows, the property of the municipalities bordering on the waters of the port, may be anchored in such places as the supervisor of anchorages may designate.

(*h*) The creation of any obstruction, not affirmatively authorized by law, to the navigable capacity of any waters, in respect of which the United States has jurisdiction, is hereby prohibited.

(*i*) No vessels shall occupy permanent berths in the anchorages except as provided for in these anchorage rules and regulations.

(*j*) The supervisor of anchorages shall assign berths in the respective anchorages to all vessels applying for them.

(*k*) All officers of revenue vessels at the port of New York are charged with the enforcement of these rules and regulations, and are empowered to remove from her anchorage any vessel not anchoring within the prescribed limits.

EAST RIVER ANCHORAGES.

(See Plate No. 1.)

1. To the northward of a line from the south point of Hart Island to Wrights Point.
2. To the westward of a line from Wrights Point to Throgs Neck.
3. To the southward of a line from buoy off Sands Point to buoy off Gangway Rock.
4. To the southward of a line from buoy off Gangway Rock to center of Stepping Stones Lighthouse.
5. To the eastward of a line from the center of Stepping Stones Lighthouse to Willets Point.
6. On Hammond Flats, to the northward of a line from Throgs Neck to Old Ferry Point.
7. To the southward of a line from Willets Point to Whitestone Point.
8. On the north side of the channel, north of a line between Old Ferry Point and Hunts Point.
9. On the south side of the channel, south of a line between Whitestone Point and buoy (No. 1) off College Point, and to the eastward of a line running from said buoy to College Point.
10. In Flushing Bay, to the southward of a line from College Point to the north end of Rikers Island.
11. To the southward of a line from the north end of Rikers Island to the north end of South Brother Island, thence to Lawrences Point.
12. To the westward of a line from Stony Point to northeast end of Wards Island; and between Wards Island and Randalls Island, and between Randalls Island and Port Morris.
13. To the westward of a line from the foot of One hundred and sixteenth street, New York, to the north end of Avenue B, New York, but no vessels shall anchor on this anchorage within 150 feet of any wharf or pier, or so as to impede the movements of a ferry, or so as to prevent ready access to or from the piers.
14. To the eastward of a line from Hatters Dock to Gibbs Point (Hallets Cove, Astoria).

(See Plate No. 2.)

15. To the southward of Thirty-second Street Pier and the northward of Twenty-fourth Street Pier, and to the westward of a line passing through the horizontal-striped buoy off Nineteenth street, running thence N. by E. ⅛ E. (cor. mag.). Vessels may anchor anywhere within these limits, provided they do not obstruct the approach to any pier or impede the movements of any ferryboat; and the officer in charge of anchorage grounds may, whenever he deems it advisable, move or cause to move any vessel not, in his opinion, complying with this proviso.

    Range for outer boundary of this anchorage: When steering on this range (N. by E. ⅛ E., cor. mag.) you should look squarely into the new Forty-second Street Ferry Slip, on the center of the axis of which is a prominent dark chimney.

SOUTH ANCHORAGE, HUDSON RIVER.

(See Plate No. 3.)

16. Vessels may anchor in the Hudson River to the westward of a line 100 yards west of the center line of said river; that the northeast boundary of this anchorage shall be a line drawn from the southeast corner of Gokey's drydock just below the Weehawken terminal of the West Shore ferries to the end of the Thirty-fourth Street Pier, New York; that the

southern limit of this anchorage be a line from the end of the Erie Railroad Company's Coal Pier, Hoboken, to the end of the Twenty-fifth Street Pier, New York, about SE. ⅜ E.

MIDDLE ANCHORAGE, HUDSON RIVER.

(SEE PLATE NO. 4.)

17. Vessels may anchor in the Hudson River within the limits of the port of New York to the westward of the center line of said river running about NE. ½ N. (cor. mag.) from Castle Point, Hoboken, through the white anchorage buoy off Sixtieth street to the northward of a line running from the southeast corner of West Shore Pier No. 3, Weehawken, to the center of the Fifty-fourth Street Pier, New York, about SE. ½ E. (cor. mag.) and to the southward of a line running from the outer end of the Guttenburg Pier to the outer end of the West Seventieth Street Pier, New York.

In no case shall a vessel anchor within 200 yards of the shore in either of these Hudson River anchorages.

UPPER ANCHORAGE, HUDSON RIVER.

(SEE PLATE NO. 4.)

18. To the northward of a line drawn from the pier on the Guttenburg side directly across the river to Eightieth street, New York (cor. mag.), to the westward of a line parallel to and 125 yards to the westward of the center line of the river.

NOTE.—Small vessels may anchor inside the pierhead lines as established by the Board of Engineers, United States Army, along the east bank of the Hudson River between Eighty-first street and One hundred and twenty-first street, and between One hundred and thirty-second street and One hundred and fifty-eighth street, in the discretion of the supervisor of anchorages, but the officer in charge of anchorage grounds may, whenever he deems it advisable, move or cause to move any vessel not, in his opinion, complying with this proviso.

NAVAL ANCHORAGE, HUDSON RIVER.

(SEE PLATE NO. 4.)

19. An anchorage is set aside for naval vessels to moor in a single line on the east side of Hudson River north of Seventy-ninth street, and thence to Fort Washington Point, and above that point if necessary. Its southernmost limit shall be north of the northernmost cable crossing the river at Seventy-ninth street, and extending northeastwardly on the east side of the river. Anchors shall be let go to the eastward of a line drawn 250 yards from the end of the pier at Seventy-ninth street to 250 yards from the end of the pier at the foot of One hundred and twenty-ninth street; thence to a point 330 yards from the pier at the foot of One hundred and fifty-eighth street; and thence northeasterly, following the general line of the 24-foot curve on the east side of the river and 250 yards distant from the salient point of this curve. No ships shall anchor within a limit of 300 yards of the prolongation of One hundred and thirtieth street, in order to give free passage for the Fort Lee ferryboats. The destroyers and other light-draft naval vessels may anchor on the west side of the river west of the 18-foot curve as shown on Coast Survey chart No. 369'.

WESTERN ANCHORAGE, UPPER BAY.

(See Plate No. 5.)

20. To the southward of a range passing through Wall Street Ferry, Brooklyn, and the white buoy to the north and east of Ellis Island; to the westward of a line running SW. by S. (nearly) from the said white buoy to a point one-half mile east from Robbins Reef Lighthouse, and to the northward of a line from Constables Point to Robbins Reef bell buoy; thence to the aforementioned point ½ mile E. of Robbins Reef Lighthouse.

In order to prevent vessels fouling the Ellis Island cable, the buoy marking the northern limit of the channel to Ellis Island and the buoy marking the southern entrance to said channel have been moved so as to leave a space of 800 yards of clear water between the anchorage grounds north of the Ellis Island Channel and the anchorage grounds south of said channel, but the ranges otherwise retain their same magnetic bearings.

No vessels shall anchor in the Black Tom or Greenville dredged channels nor near the entrances to said channels so as to obstruct the approaches or interfere in any way with the free navigation of the same.

Note.—Vessels are especially cautioned not to anchor in Ellis Island Channel, thereby endangering the cable in said channel. In addition to the penalty for illegal anchorage, the owners of vessels which foul the above-mentioned cable will be liable for the damage resulting therefrom, including the cost of clearing, which should be done, in order to reduce the injury to a minimum, by signaling for the Western Union Company's tug.

EASTERN ANCHORAGE, UPPER BAY.

(See Plate No. 6.)

21. To the southward of a line passing through the Statue of Liberty on Bedloes Island, the two white buoys marking the north limit of anchorage ground and the southern point of the north entrance to the Erie Basin; to the eastward of a range passing through Produce Exchange tower and buoy No. 14 and bell buoy off Owls Head, and thence marked on the eastern and southeastern limits by four white anchorage buoys along the western edge of the widened Bay Ridge and Red Hook channels. These buoys will eventually be replaced by proper channel buoys marking the edge of the dredged channel. On and after July 1, 1901, vessels will not be allowed to anchor to the eastward and southward of said line of buoys. Small vessels may, in the discretion of the supervisor of anchorages, anchor to the southward and eastward of the Bay Ridge Channel, provided they are inside of the pierhead lines as established by the Board of Engineers, United States Army. Small vessels may, in the discretion of the supervisor of anchorages, anchor at the mouth of Gowanus Bay, to the eastward of a line tangent to the southwestern edge of Erie Basin Bulkhead, and running thence S. by E. (cor. mag.), but so as to leave a clear channel of 150 yards along the northern shore. The supervisor of anchorages may, in his discretion, remove any vessel not complying with the provisions hereof.

EASTERN ANCHORAGE, LOWER BAY.

(See Plate No. 7.)

22. To the eastward of a line drawn through Fort Lafayette and buoy No. 4 (approximate magnetic bearing S. ⅞ E.) of Ambrose Channel, and to the northward of such buoys as may ultimately mark the new Ambrose Channel, except that to the southward of Nortons

Point and Coney Island, vessels shall not anchor on or to the northward of the line of the Commercial Cable, as shown on the charts, thus insuring a free passage to shipping using the new Ambrose Channel or bound to the eastward along the south shore of Coney Island.

STATEN ISLAND ANCHORAGE.

(SEE PLATES NOS. 7 and 8.)

23. To the southward of a line from St. George Ferry Flagstaff to the white buoy off St. George Landing, and to the westward of a line running S. $\frac{5}{8}$ W. (nearly) from the white buoy off St. George Landing, through the white buoy off Tompkinsville, and as far south as the white buoy off Clifton, Staten Island. (See Plate No. 8.) To the westward of a line running SSE. $\frac{1}{4}$ E. (nearly) from Fort Tompkins to the buoy on Craven Shoal; thence to buoys Nos. 11, 9, and 7; thence to Conovers Beacon.

The part of anchorage 23 lying between the northern boundary and the white buoy 800 yards south of said boundary is reserved for ships of war of all nations and vessels of the United States Government.

QUARANTINE ANCHORAGE.

(SEE PLATE NO. 8.)

24. To the southward of a line passing through Clifton, Staten Island, and the white buoy off this point, and to the westward of a line from the buoy off Clifton, Staten Island, to the bell at Fort Wadsworth.

Vessels arriving at quarantine and awaiting inspection may anchor temporarily to the westward of a range passing through Craven Shoal Buoy and Robbins Reef Lighthouse, but as soon as cleared by the quarantine officer must vacate this temporary anchorage, and if detained in quarantine, must at once move into the quarantine anchorage.

SANDY HOOK ANCHORAGE.

(SEE PLATE NO. 7.)

25. To the southward of a line extending from East Beacon to Bayside Beacon (Point Comfort). In order to prevent injury to the submarine cables, vessels are forbidden to anchor when the East Beacon Lighthouse bears anywhere between the compass bearings of WSW. $\frac{1}{2}$ W. and SW. by W. $\frac{3}{4}$ W. from the vessel, unless the said vessel is to the northward of the northern line of buoys of Gedneys Channel.

(SEE PLATE NO. 7.)

26. Vessels may anchor on Dry Romer Shoal and Flynns Knoll. Care must be observed not to foul the United States Signal Corps cable (shown in red on the chart), which crosses Flynns Knoll.

ANCHORAGE FOR EXPLOSIVES.

(SEE PLATES NOS. 1 and 5.)

27. Vessels carrying gunpowder or other explosives may anchor only as follows:

First. On the shoal ground to the eastward of Rikers Island, East River, from one-fourth to five-eighths of a mile from this island.

Second. On the New Jersey Flats between a line drawn parallel to and 1,500 feet to the south of the Black Tom dredged channel and a line drawn parallel to and 1,500 feet north of the Greenville dredged channel, and to the westward of a line from Bedloes Island to Robbins Reef, provided that they do not anchor within 1,000 yards of Bedloes Island, or within 500 yards of any pier.

Third. Vessels (carrying explosives) of too great draft to use the above anchorages may anchor only in Gravesend Bay, on a line drawn from Fort Hamilton to the western tip of Nortons Point, Coney Island, but not within 1,000 yards of the shore. All vessels laden with explosives while within the port will display at all times a red flag of at least 16 square feet surface at the masthead. Vessels so laden and without masts will display the flag at least 10 feet above the uppermost deck. All such vessels must be at all times in charge of competent persons and must comply with the navigation laws in regard to lights and fog signals.

Oscar S. Straus,
Secretary.

PLATE I

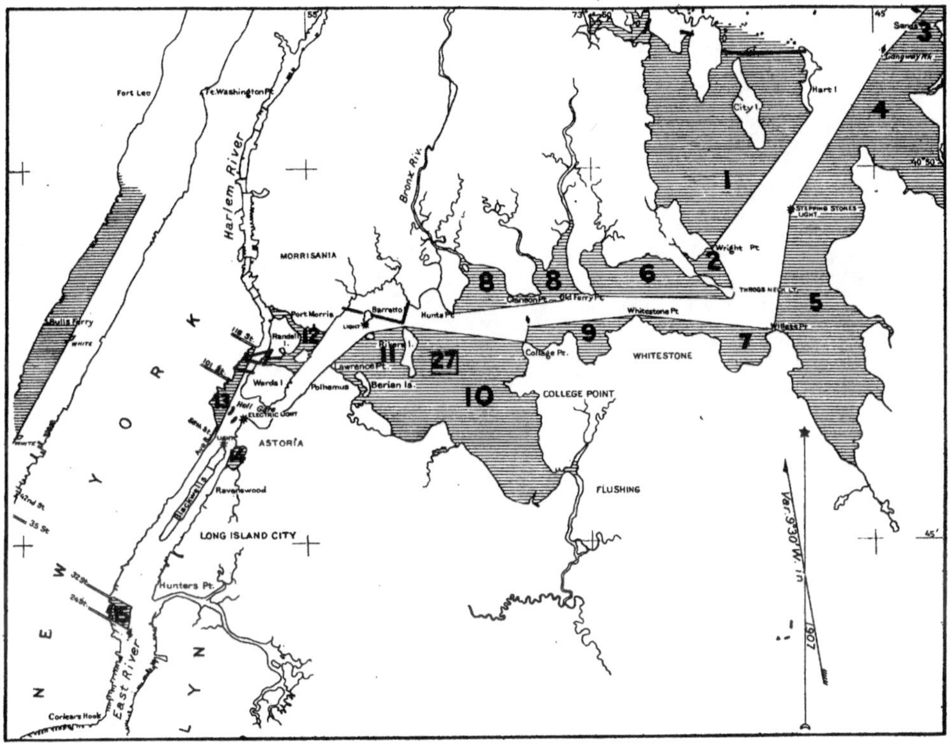

EAST RIVER ANCHORAGES.

CABLE CROSSING ON ANCHORAGE GROUNDS SHOWN THUS

ANCHORAGE THUS

PLATE II

EAST RIVER ANCHORAGE OFF 26TH STREET
BEARINGS ARE MAGNETIC
SCALE OF YARDS
100 50 0 100 200 300 400 500
NAUTICAL MILE
44TH ST.
43RD ST.
42ND ST
41ST ST.
40TH ST.
39TH ST
38TH ST.
37TH ST.
36TH ST.
35TH ST
34TH ST.
33RD ST.
32ND ST.
31ST ST.
30TH ST.
29TH ST.
28TH ST.
27TH
26TH ST.
25TH ST.
24TH ST.
23RD ST.
22ND ST.
21ST ST.
20TH ST.
19TH ST.
18TH ST.
FIRST AVE.
AVE. A.
AVE.
LONG ID CITY FERRY
N.W. ½ W.
N by E. ⅛ E.
15
N.W. ½ W.
GREENPOINT FERRY
FERRY TO BROADWAY, BOOKLYN
LONG ISLAND R.R. FERRY SLIPS
DOCK ST.
PIER ST.
NEWTOWN CREEK
VAR'N 9°30' W. IN 1907
ANNUAL INCREASE 3'
GREEN ST.
10TH & 23RD ST FERRIES
HORIZONTAL STRIPED BUOY
ANCHORAGE THUS

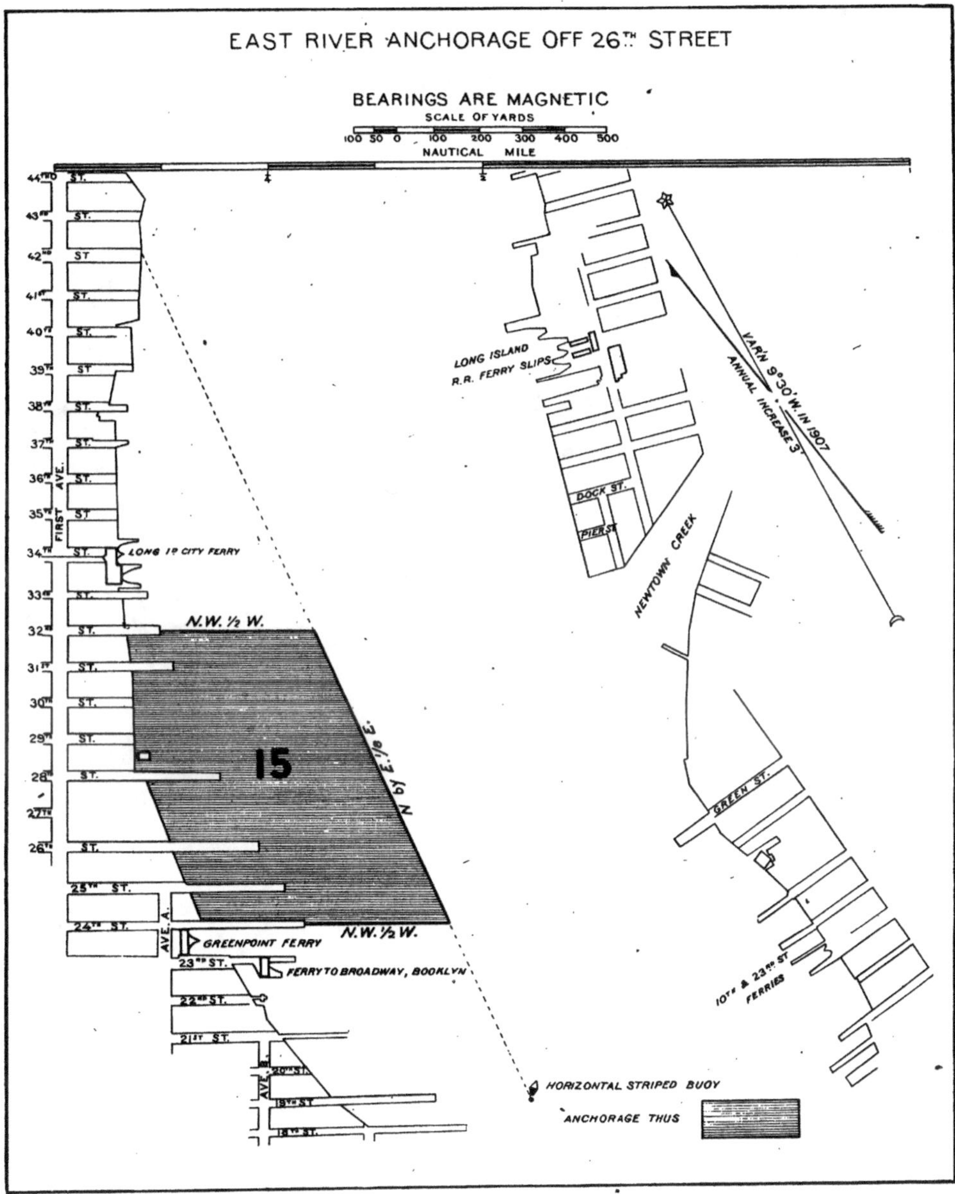

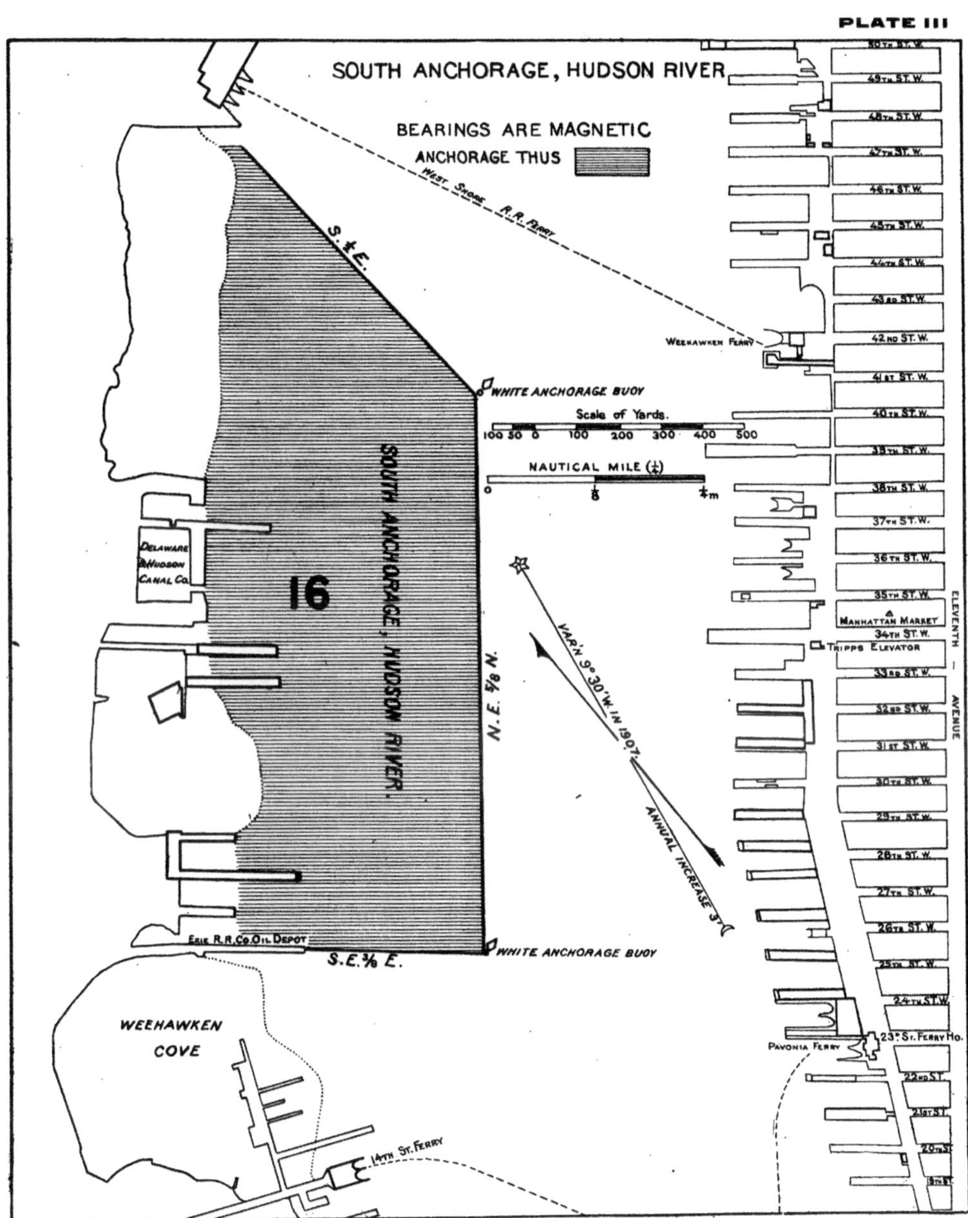
SOUTH ANCHORAGE, HUDSON RIVER
BEARINGS ARE MAGNETIC
ANCHORAGE THUS
WEST SHORE R.R. FERRY
S. ¼ E.
WHITE ANCHORAGE BUOY
Scale of Yards.
100 50 0 100 200 300 400 500
NAUTICAL MILE (¼)
0 ⅛ ¼m
SOUTH ANCHORAGE, HUDSON RIVER.
16
N. E. ⅝ N.
DELAWARE & HUDSON CANAL CO.
VAR'N. 9° 30' W. IN 1907.
ANNUAL INCREASE 3'
ERIE R.R. CO. OIL DEPOT
S.E. ⅜ E.
WHITE ANCHORAGE BUOY
WEEHAWKEN COVE
14TH ST. FERRY
WEEHAWKEN FERRY
PAVONIA FERRY
23RD ST. FERRY HO.
MANHATTAN MARKET
TRIPPS ELEVATOR
ELEVENTH - AVENUE
50TH ST. W.
49TH ST. W.
48TH ST. W.
47TH ST. W.
46TH ST. W.
45TH ST. W.
44TH ST. W.
43RD ST. W.
42ND ST. W.
41ST ST. W.
40TH ST. W.
39TH ST. W.
38TH ST. W.
37TH ST. W.
36TH ST. W.
35TH ST. W.
34TH ST. W.
33RD ST. W.
32ND ST. W.
31ST ST. W.
30TH ST. W.
29TH ST. W.
28TH ST. W.
27TH ST. W.
26TH ST. W.
25TH ST. W.
24TH ST. W.
22ND ST.
21ST ST.
20TH ST.
19TH ST.

PLATE IV

UPPER, NAVAL AND MIDDLE ANCHORAGES
HUDSON RIVER
BEARINGS ARE MAGNETIC
ANCHORAGE THUS
CABLE CROSSING ON ANCHORAGE GROUND THUS
Scale of Yards
100 50 0 100 200 300 400 500
NAUTICAL MILE (½)
FT. WASHINGTON
18
19
17
Upper Anchorage, Hudson River.
Naval Anchorage, Hudson River
Middle Anchorage, Hudson R.
NEW JERSEY
NEW YORK CITY
130 ST. W
110 ST. W
100 ST. W
90 ST. W.
80 ST. W
70 ST. W
60 ST. W
WHITE A
Days Pt.
VAR N 9° 30' WEST IN 1900
ANNUAL INCREASE, 3'

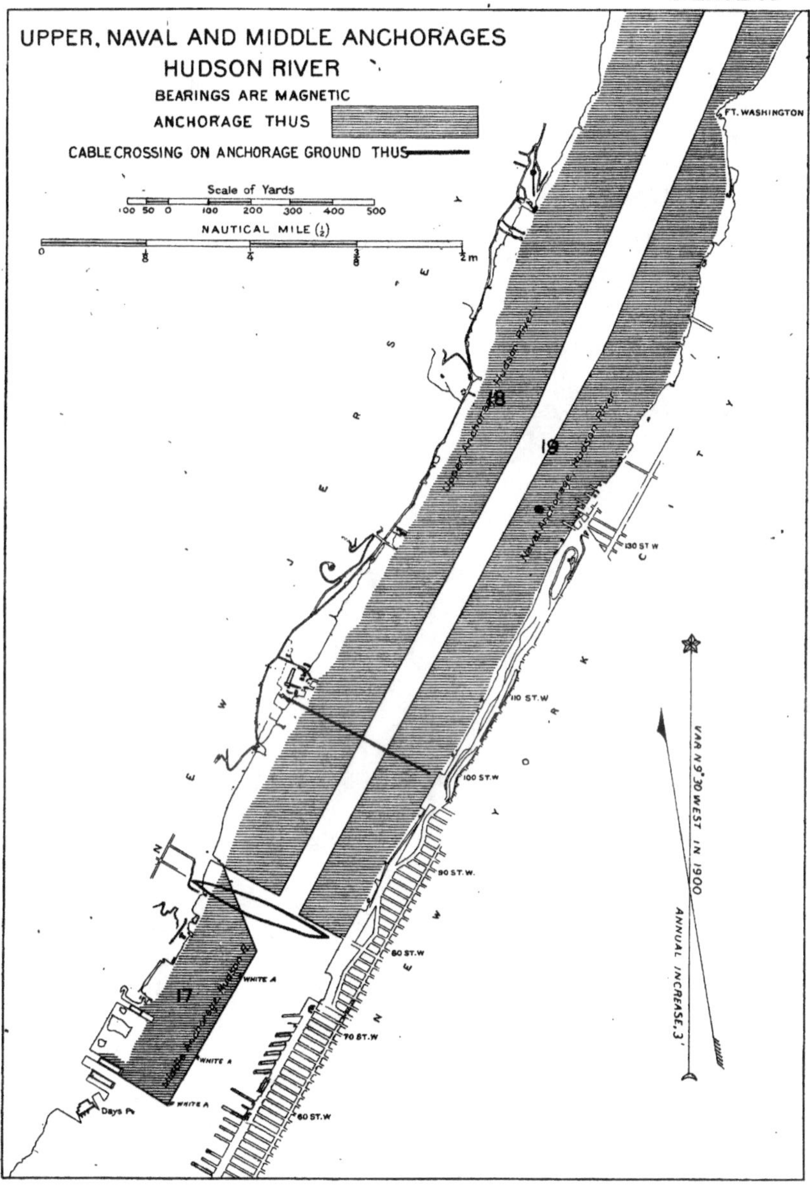

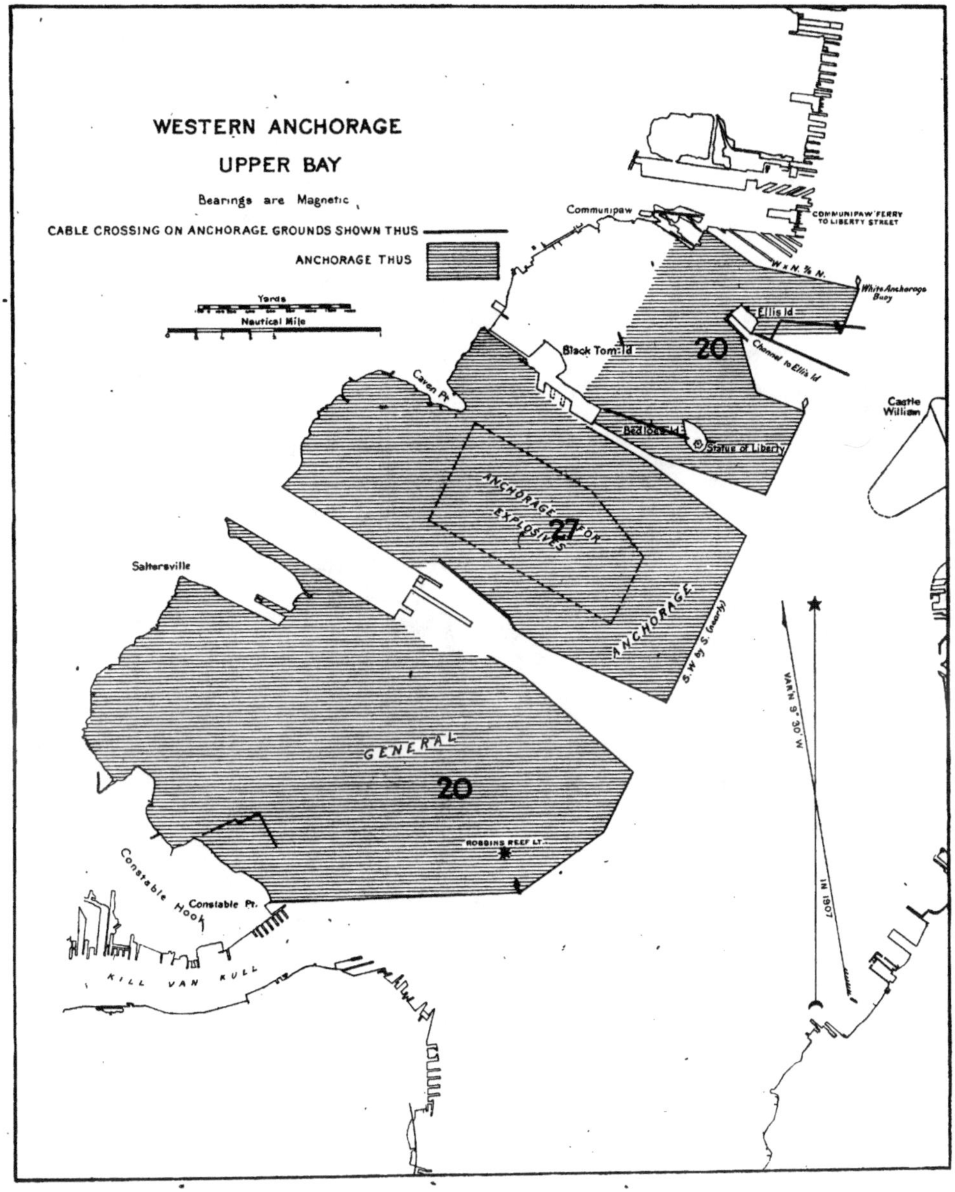

WESTERN ANCHORAGE UPPER BAY

Bearings are Magnetic

CABLE CROSSING ON ANCHORAGE GROUNDS SHOWN THUS

ANCHORAGE THUS

Yards

Nautical Mile

Communipaw

COMMUNIPAW FERRY TO LIBERTY STREET

W x N. ¾ N.

White Anchorage Buoy

Ellis Id

Channel to Ellis Id

Black Tom Id

20

Caven Pt

Castle William

Bedloes Id

Statue of Liberty

ANCHORAGE FOR EXPLOSIVES

27

ANCHORAGE

S.W. by S. (nearly)

Saltersville

GENERAL

20

ROBBINS REEF LT.

VAR'N 9° 30' W IN 1907

Constable Hook

Constable Pt.

KILL VAN KULL

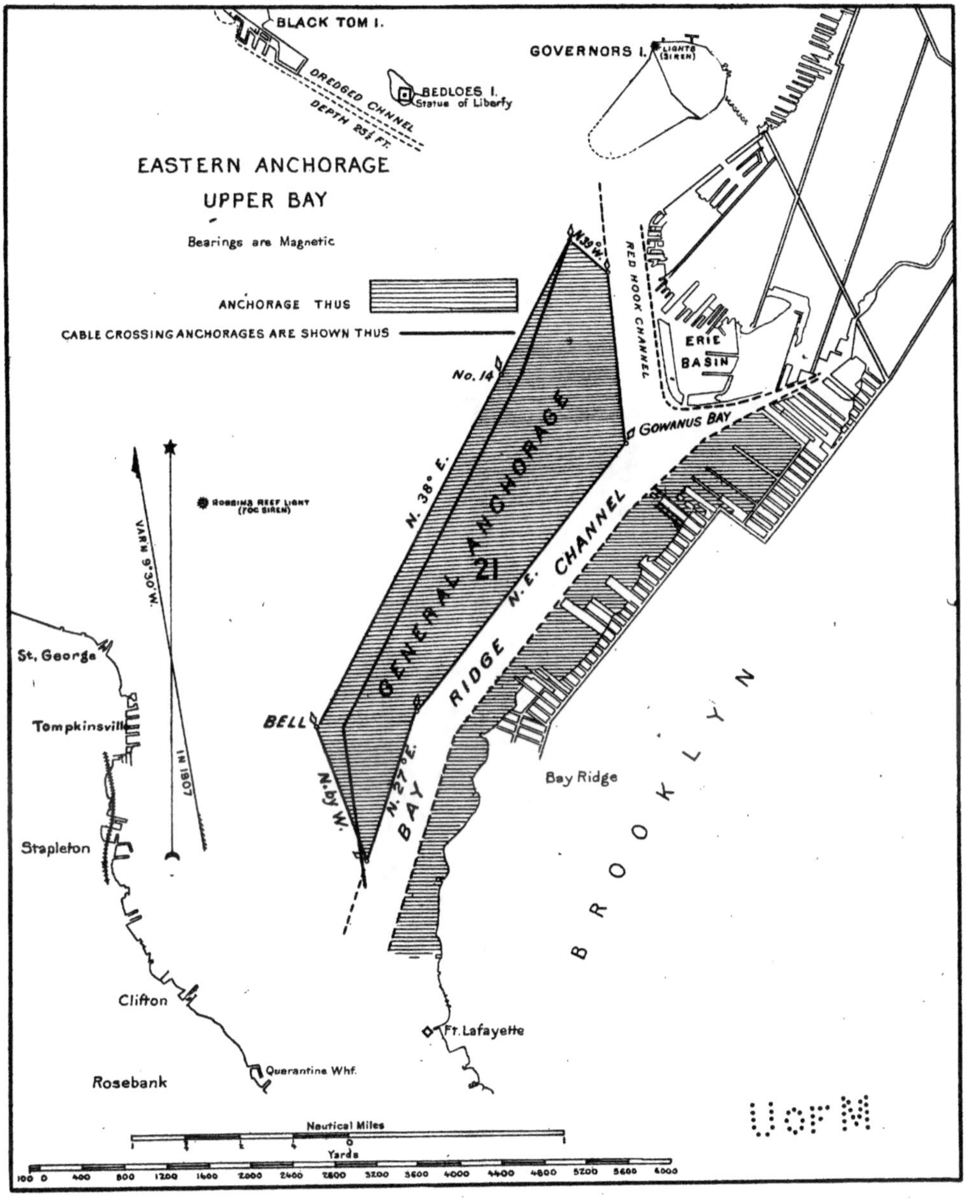

PLATE VI

EASTERN ANCHORAGE
UPPER BAY

Bearings are Magnetic

ANCHORAGE THUS

CABLE CROSSING ANCHORAGES ARE SHOWN THUS

BLACK TOM I.

DREDGED CHANNEL DEPTH 25½ FT.

BEDLOES I.
Statue of Liberty

GOVERNORS I.

LIGHTS (SIREN)

RED HOOK CHANNEL

ERIE BASIN

N.39°W.

No. 14

GOWANUS BAY

GENERAL ANCHORAGE 21

N. 38° E.

ROBBINS REEF LIGHT (FOG SIREN)

VAR'N 9°30'W. IN 1907

BAY RIDGE N.E. CHANNEL

BELL

N.by W.

N.27°E.

BAY RIDGE

St. George

Tompkinsville

Stapleton

Clifton

Rosebank

Quarantine Whf.

Ft. Lafayette

Bay Ridge

BROOKLYN

Nautical Miles

Yards

100 0 400 800 1200 1600 2000 2400 2800 3200 3600 4000 4400 4800 5200 5600 6000

PLATE VII

LOWER BAY AND SANDY HOOK ANCHORAGES

CABLE CROSSING ON ANCHORAGE GROUNDS SHOWN THUS

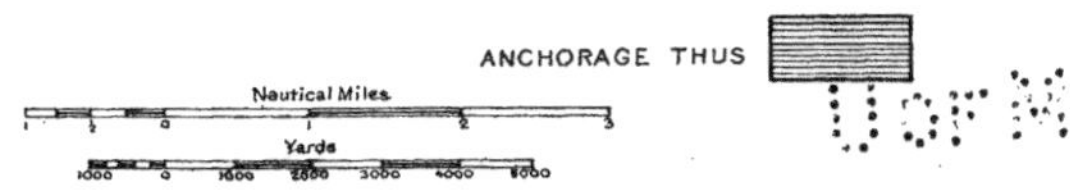

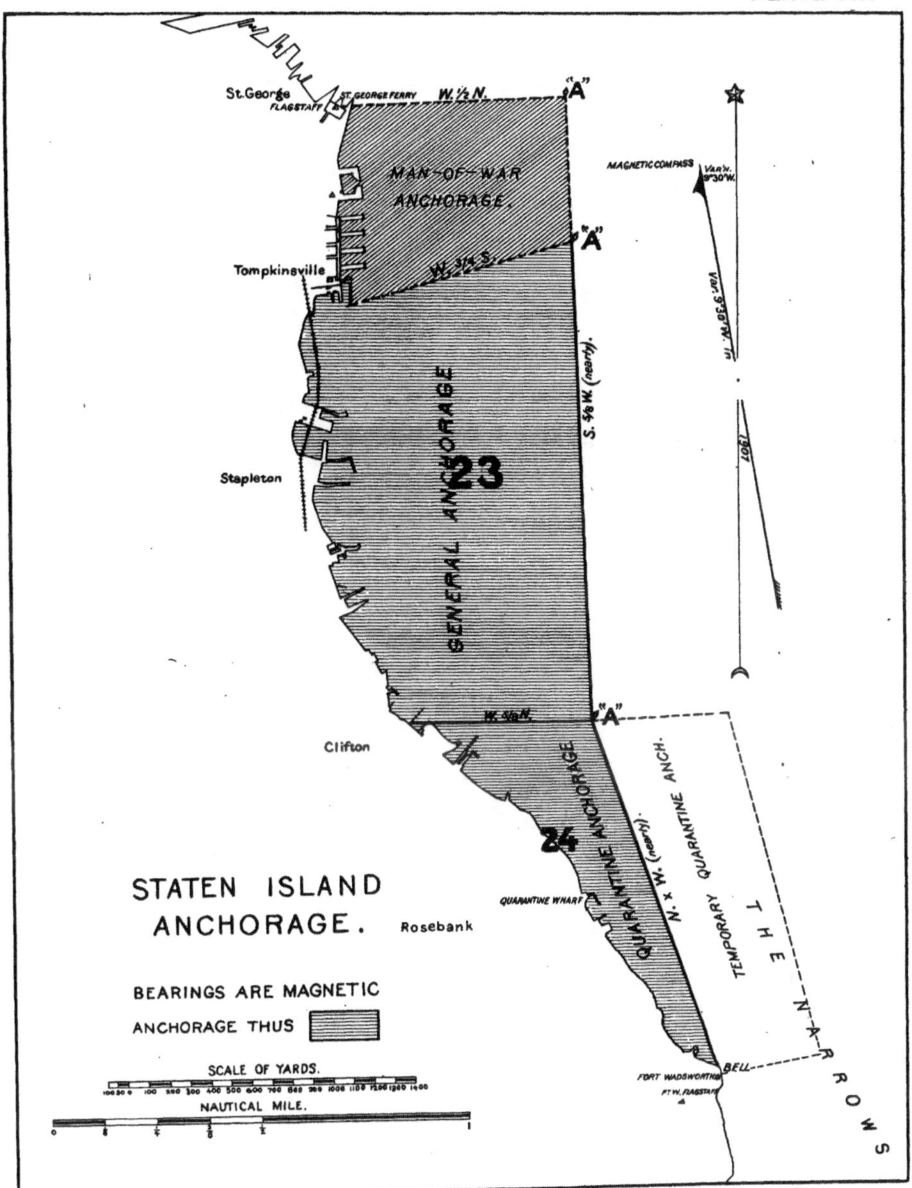
St. George
FLAGSTAFF
ST. GEORGE FERRY
W. ½ N.
"A"
MAN-OF-WAR ANCHORAGE.
MAGNETIC COMPASS
Var'n. 9°30' W.
Var. 9°35' W. in 1907
Tompkinsville
W. 3/4 S.
"A"
S. 5/8 W. (nearly).
GENERAL ANCHORAGE
23
Stapleton
W. 5/8 N.
"A"
Clifton
QUARANTINE ANCHORAGE
24
N. x W. (nearly).
TEMPORARY QUARANTINE ANCH.
THE NARROWS
QUARANTINE WHARF
STATEN ISLAND ANCHORAGE.
Rosebank
BEARINGS ARE MAGNETIC
ANCHORAGE THUS
SCALE OF YARDS.
NAUTICAL MILE.
BELL
FORT WADSWORTH
FT. W. FLAGSTAFF

www.ingramcontent.com/pod-product-compliance
Lightning Source LLC
LaVergne TN
LVHW020631110826
845149LV00004B/1142

* 9 7 8 1 4 1 8 1 8 6 6 8 5 *